Studies in Spermatogenesis

(Part 2)

N. M. Stevens

Alpha Editions

This edition published in 2024

ISBN : 9789364734547

Design and Setting By
Alpha Editions
www.alphaedis.com
Email - info@alphaedis.com

As per information held with us this book is in Public Domain.
This book is a reproduction of an important historical work. Alpha Editions uses the best technology to reproduce historical work in the same manner it was first published to preserve its original nature. Any marks or number seen are left intentionally to preserve its true form.

Contents

INTRODUCTION. ....- 1 -

COLEOPTERA...- 2 -

HEMIPTERA HOMOPTERA..- 13 -

LEPIDOPTERA...- 16 -

SUMMARY OF RESULTS. ....- 17 -

COMPARISON OF RESULTS IN DIFFERENT SPECIES OF COLEOPTERA...- 21 -

COMPARISON OF THE COLEOPTERA WITH THE HEMIPTERA AND LEPIDOPTERA. ....- 23 -

GENERAL DISCUSSION. ....- 24 -

BIBLIOGRAPHY..- 28 -

DESCRIPTION OF PLATES..- 30 -

INTRODUCTION.

In Part I of this series of papers, the spermatogenesis of five species belonging to four different orders of insects was considered. In two species of Orthoptera an "accessory chromosome" was found; in *Tenebrio molitor*, one of the Coleoptera, an unequal pair of chromosomes was described; in the other species no heterochromosomes were discovered. The apparent bearing of the chromosome conditions in *Tenebrio molitor* on the problem of sex determination has led to a further investigation of the germ cells of the Coleoptera. One of the Hemiptera homoptera and two of the Lepidoptera have also been examined for comparison with the Coleoptera and the Hemiptera heteroptera.

METHODS.

As a result of previous experience with similar material, only two general methods of fixing and staining have been employed: (1) Fixation in Flemming's strong solution or Hermann's platino-aceto-osmic, followed by either Heidenhain's iron-hæmatoxylin or Hermann's safranin-gentian staining method (Arch. f. mikr. Anat. 1889). (2) Fixation after Gilson's mercuro-nitric formula, followed by iron-hæmatoxylin, Delafield's hæmatoxylin and orange G, Auerbach's combination of methyl green and acid fuchsin, or thionin.

The iron-hæmatoxylin with either mode of fixation gives by far the most satisfactory preparations for general study. The other stains were used mainly for the purpose of distinguishing between heterochromosomes and plasmosomes in resting stages of the nucleus.

COLEOPTERA.

Trirhabda virgata (Family Chrysomelidæ).

Two species of *Trirhabda* were found in larval, pupal, and adult stage on *Solidago sempervirens*, one at Harpswell, Maine, the other at Woods Hole, Massachusetts. The adult insects of the two species differ slightly in size and color, the germ cells mainly in the number of chromosomes, *Trirhabda virgata* having 28 and *Trirhabda canadense* 30 in spermatogonia and somatic cells.

In *Trirhabda virgata*, the metaphase of a spermatogonial mitosis (plate VIII, fig. 3) contains 28 chromosomes, one of which, as in *Tenebrio molitor* is very much smaller than any of the others. The maternal homologue of the small chromosome is, as later stages show, one of the largest chromosomes. In *Tenebrio* the unequal pair could not be distinguished in the growth stages of the spermatocytes. In *Trirhabda* it has not been detected in the synizesis stage (fig. 4), but in the later growth stages (figs. 5-7) this pair is conspicuous in preparations stained by the various methods cited above, while the spireme is pale and inconspicuous. The size of the heterochromosome pair varies considerably at different times in the growth period, and in some nuclei (fig. 7) both chromosomes appear to be attached to a plasmosome. The ordinary chromosomes assume the form of rings and crosses in the prophase of the first maturation mitosis (fig. 8), but usually appear in the spindle as dumb-bells or occasionally as tetrads (fig. 10), or crosses (fig. 11). The unsymmetrical pair is plainly seen in figures 9 and 11, but is not distinguishable in a polar view of the metaphase (fig. 13). In the anaphase (figs. 14-16) the larger and the smaller components of the pair separate as in *Tenebrio*. This is, therefore, clearly a reducing division as far as this pair is concerned, and probably for all of the other pairs, though neither the synapsis stage nor the prophase forms are so clear on this point as in some of the other species studied. Figures 17 and 18 show metaphases of the two classes of second spermatocytes, the chromosomes varying somewhat in form in different preparations and even in different cysts of the same preparation. An early anaphase of this mitosis is shown in figure 19; here the small chromosome is already divided. It was impossible to find good polar views of the daughter plates in the two classes of second spermatocytes, but it is evident from figure 19 and other similar views of the second spermatocyte spindle that, as in *Tenebrio*, one-half of the spermatids will contain one of the derivatives of the small chromosome, the other half one of the products of its larger homologue.

Sections of male pupæ were examined for equatorial plates of somatic mitoses. Figure 1 is a specimen of such plates. As might be expected, this figure resembles quite closely the spermatogonial equatorial plate (fig. 3) in number, form, and size of chromosomes, the small one being present in both. Figure 2 is from the follicle of a young egg; here we find 28 chromosomes, but no small one. The chromosome corresponding to the larger member of the unequal pair in the male evidently has a homologue of equal size in the female. The chromosome relations in the male and female somatic cells are therefore the same as in *Tenebrio molitor*, and must have been brought about by the development of a male from an egg fertilized by a spermatozoön containing the small chromosome, and a female from an egg fertilized by a spermatozoön containing the larger heterochromosome.

Trirhabda canadense.

In *Trirhabda canadense* the spermatogonial chromosomes are invariably smaller than in *T. virgata*, but similar size relations prevail. The spermatogonial plate (fig. 21) contains 30 chromosomes, 29 large and 1 extremely small. In the growth stages the association of the two unequally paired chromosomes with a rather large plasmosome is more evident than in *T. virgata* (figs. 22-23). In this species the unequal pair is more often found at a different level from the other chromosomes in the early metaphase of the first maturation mitosis (fig. 24), but it later comes into the plate with the other chromosomes (figs. 25-27), and divides earlier than most of the other bivalents (fig. 27). In a polar view of this metaphase the largest chromosome often appears double (fig. 28); in a front view it is a tetrad as in *T. virgata*, figure 10. Figure 29 is the equatorial plate of a metaphase in which the larger component of the unequal pair has been removed in sectioning. The daughter plates of a first spermatocyte in anaphase (fig. 30) show the separation of the components of the heterochromosome pair; and equatorial plates of the resulting two classes of second spermatocytes (fig. 31) show the same conditions. Figures 32 and 33 are prophases of the second division, figure 33 showing the small chromosome ready for metakinesis. It was impossible here also to get good drawings of daughter plates of the second spermatocytes to show the content of the two classes of spermatozoa, but there is no doubt that all of the chromosomes divide in the second mitosis, giving one class of spermatids containing the small chromosome, the other class its larger homologue.

No male somatic cells were found in mitosis, but they would, if found, show the same conditions as in the spermatogonia. One of many good equatorial plates from egg follicles (fig. 20) shows 30 large chromosomes, indicating an equal pair in place of the unequal pair of the male.

Chelymorpha argus (Family Chrysomelidæ).

This species was found in larval and adult stages on *Convolvulus arvensis* at Harpswell, Maine, in July and August. It shows the same conditions as *Trirhabda* and *Tenebrio*, so far as the unequal pair of chromosomes is concerned, and is especially favorable for study of synapsis stages. The number of chromosomes in the spermatogonia (plate IX, fig. 36) is 22. Here the components of the unequal pair are the small spherical chromosome and one of the several chromosomes third in size, forming a comparatively small unsymmetrical bivalent (figs. 47-49). The spermatogonia occupy the outer end of each follicle, and next to them comes a layer of cysts in which the chromosomes from the last spermatogonial division are closely massed in the form of short deeply staining loops at one side of the nuclear space (fig. 37). Following this synizesis stage comes one in which some of the short loops have straightened, their free ends extending out into the nuclear space (figs. 38 and 39). Figure 40 shows the nucleus of a slightly later stage in which the free ends of two straightened chromosomes are on the point of uniting. In figures 41 and 42 the point of union of homologous chromosomes is indicated in some cases by a knob, in others by a sharply acute angle. In a slightly later stage (fig. 43), when all of the short loops have straightened and united in pairs, the point of union is no longer visible, all of the loops being rounded at the bend and of equal thickness throughout. My attention was first called to this method of synapsis by the conspicuous difference in number and length of loops in the synizesis stage compared with the later bouquet stage just before the spireme is formed. Following the synapsis stage shown in figure 43 comes one in which the loops lose their polarized arrangement and unite to form a continuous spireme (figs. 44 and 45). In this form, the heterochromosome pair could not be distinguished until the spireme stage, and it is, therefore, uncertain whether these chromosomes remain condensed after the last spermatogonial divisions and are hidden among the massed and deeply staining loops of the synizesis and synapsis stages, or whether they pass through the same synaptic phases as the other chromosomes, condensing and remaining isolated at the beginning of the spireme stage. An early prophase of the first maturation mitosis (fig. 46) shows segments of the spireme longitudinally split, and in some cases transformed into crosses which show a transverse division also. Most of the equal bivalents have the dumb-bell form in the spindle (figs. 47-49). One is ring-shaped, the ring being formed by union of the free ends of the segment so that the spindle fibers are attached to the middle of each univalent chromosome (fig. 49). This method of ring formation, like that described by Montgomery ('03) for the Amphibia, is of very frequent occurrence in the spermatocytes of the Coleoptera. The dumb-bells are so bent at the ends (fig. 52) that the spindle fibers, here also, are attached at or near the center of each univalent component of a bivalent chromosome, and the separated, univalent chromosomes go to the poles of the spindle in the form of Vs. As in *Tenebrio* the heterochromosome pair is late about coming into the equatorial plate (figs. 47-48), but it does finally take its position with the others (fig. 49) and separates into its component parts somewhat earlier than the other bivalents (figs. 52, 53). Figures 50 and 51 show polar views of the metaphase, the smaller element (x) being the unequal pair. The chromosomes in late anaphase are too much crowded to give clear drawings. As in all the beetles so far studied there is no rest stage between the two maturation divisions, but the late anaphase of the first mitosis passes over quickly into the second spindle. Figures 54 and 55 are typical equatorial plates of the second division, one showing the small chromosome (s), the other its mate more nearly spherical than the others (l). An anaphase including the small chromosome is shown in figure 56. As in the species previously described the spermatozoa are evidently dimorphic.

Female somatic equatorial plates from egg follicles are shown in figures 34 and 35; 22 chromosomes are present and no one is without an equal mate.

Odontota dorsalis (Family Chrysomelidæ).

Odontota dorsalis is a small leaf-beetle found on *Robinia pseudacacia*. The chromosomes are comparatively few in number, 16 in the spermatogonia (figs. 58 and 59), and of immense size when one considers the smallness of the beetle. In some of the spermatogonial cysts many of the chromosomes are V-shaped as in figure 58, while in others all, with the exception of the small one, are rod-shaped as in figure 59, which looks like a hemipteran equatorial plate. The spermatogonial resting nucleus (fig. 60) contains a large plasmosome (p), but no condensed chromatin. The synizesis and synapsis stages are similar to those in Chelymorpha (figs. 61 and 62). The spireme stage (figs. 63, 64) contains, in addition to the pale spireme, a very conspicuous group consisting of a large plasmosome with a large and a small chromosome attached to it. In the prophase, before the nuclear membrane has disappeared, this group is easily distinguished from the other dumb-bell and ring-shaped bivalents (figs. 65-67). In preparations much destained (fig. 67) the small chromosome component of the group retains the stain longer than the larger one. The spindle in prophase (fig. 68) is much elongated and the 8 chromosomes are often spread out upon it so as to be easily counted. In the early metaphase the parachute-like heterochromosome group is always nearer one pole of the spindle (plate X, figs. 69 and 70). The equatorial plate often shows both the larger component of the pair and the plasmosome (fig. 71). Figures 72-74 show the metakinesis of the heterochromosome bivalent. In figure 74 the two unequal elements are completely separated and the plasmosome has disappeared. The equatorial plates of the two resulting kinds of second spermatocytes appear in figures 75 and 76. In the anaphase of the second division all of the chromosomes are divided quantitatively as may be seen in figures 77 and 78. A few dividing male somatic cells were found in the walls of the testis. Figure 57 (plate IX) is an equatorial plate from one of these. The chromosomes are like those of the spermatogonia (figs. 58 and 59), 15 large and 1 small. No dividing female somatic cells were found.

A few drawings of developing spermatids are given to show the transformations of a peculiar body which seems to be characteristic of insect spermatids. Figure 79 is a very young spermatid showing only diffuse chromatin in the nucleus. The nucleus soon enlarges (fig. 80) and a large dense body (*n*) appears which stains like chromatin with various staining media. A little later (fig. 81) the chromatin forms a homogeneous, more or less hemispherical or sometimes crescent-shaped mass which stains an even gray in iron-hæmatoxylin. In addition the nucleus contains a body (*n*) smaller than in the preceding stage, but staining the same. As the nucleus condenses and elongates to form the sperm head, a light region containing this deeply staining body is seen on one side (figs. 82, 83). A little later the body is divided into two, which appear sometimes spherical (fig. 84), sometimes elongated (fig. 85). As the sperm head elongates still more, approaching maturity, these bodies diminish in size (figs. 86, 87) and ultimately disappear. A cross section of the sperm head at such a stage as figure 87 shows the chromatin in crescent shape with material which stains very little within (fig. 88). The chromatin-like body described above was observed in *Tenebrio* in a stage corresponding to figure 81, and it was thought that the larger body seen in some cases and the smaller one in others might be the larger and smaller heterochromosomes, but a study of this element in more favorable material disproves that supposition by showing that the different sizes are merely different phases in the evolution of the body. Throughout its history it stains like dense chromatin, and my only suggestion as to its origin is that it seems, from a study of this and other species of beetles, to be a derivative of the chromatin of the spermatid, increasing in size for a time, then decreasing, and finally breaking up into granules and dissolving in the karyolymph. Whether it has any function connected with the development of the spermatozoön, or whether it is merely material rejected from the chromosomes, as in many cases in oögenesis, one can only surmise.

In one testis a peculiar abnormality was found. In all of the perfect spermatogonial plates two small chromosomes were present (figs. 89, 90). Nineteen such plates were counted in five different cysts. All of the equatorial plates of the first spermatocytes showed 8 chromosomes, as usual. In a few favorable growth stages (fig. 91) the two small chromosomes were seen to be combined with the larger heterochromosome and a plasmosome, and one first spermatocyte spindle was found in which the same combination could be clearly seen (fig. 92). All of the second spermatocyte metaphases in which a small chromosome occurred, contained two small ones, making 9 in all (fig. 93). The others contained 8 large chromosomes, as usual. The only explanation suggested by the conditions is that somewhere in its history, the small chromosome had undergone an extra division, and that ever afterward the two products behaved like the one small heterochromosome of a normal individual. The chief interest in this abnormality centers in the fact that the two small chromosomes of this specimen behave exactly like the usual single one, emphasizing the individuality of this particular heterochromosome. Both evidently have the same individual characteristics and affinities as the one in other cases.

Epilachna borealis (Family Coccinellidæ).

Epilachna borealis was found in abundance on squash vines at Woods Hole, Massachusetts, in September. The testes, unlike those of most of the Coleoptera, consist of many free follicles similar to those of the Orthoptera. The germ glands were rather far advanced, but some good spermatogonial and spermatocyte cysts were found. In figure 94, a spermatogonial metaphase, the small chromosome is shown with 17 larger ones. The heterochromosome pair appears in condensed form in the spireme stage (fig. 95), and again in the first maturation spindle (figs. 96, 97). The varying forms of the ordinary chromosomes are shown in figure 98. Figures 99 and 100 are equatorial plates of the first mitosis. The unequal pair is shown by itself in figure 101, and the separation of the heterochromosomes is seen in figure 102. Equatorial plates of the second division, one containing the small chromosome (*b*), are shown in figure 103. A prophase of the same division (fig. 104) proves that the small chromosome divides quantitatively like the others. It was interesting to find here and there in this material whole cysts in which the nuclei were like those described by Paulmier ('99) for *Anasa tristis* (plate XIII, fig. 14) as cells which were being transformed to serve as food for the glowing spermatids (figs. 105, 106). The only occasional appearance of these cysts seems to me to preclude their being a special dispensation to furnish the spermatids with nutrition during their transformation. Their appearance and size make me suspect that they are giant spermatids due to the failure of one of the spermatogonial or spermatocyte mitoses. The smaller chromatin body seems to correspond to that described for the spermatids of *Odontota dorsalis*.

Euphoria inda (Family Scarabæidæ).

Of *Euphoria inda* only one male was captured, but the numerous testes furnished abundant material in desirable stages. The spermatogonial equatorial plate (fig. 107) contains 20 chromosomes of which the two smallest (*l* and *s*) form the unequal pair. The resting spermatogonium contains a two-lobed plasmosome (fig. 108). The growth stages are similar to those in *Tenebrio* in showing no distinct bouquet stage, but there is a spireme stage in which the heterochromosome pair is clearly seen (fig. 109). Figure 110 (plate XI) is an early prophase, and figure 111 one in which the unequal pair appears with a tetrad and several dumb-bell forms. The prophase of the spindle, as in *Odontota*, is much elongated (fig. 112). In figures 113-116 the small heterochromosome pair is shown in various positions with reference to the other chromosomes of the metaphase of the first spermatocyte. Figure 117 shows it more deeply stained than the others in the equatorial plate. This pair divides in advance of the others, and the larger and smaller elements are plainly seen nearer the poles in anaphase than the other univalent chromosomes (figs. 118-120). Daughter plates of the first spermatocyte are shown in figure 121, and equatorial plates of the second spermatocyte in figure 123. Figure 122 shows the telophase of the first division with the spindle for the second division forming. In figures 124 and 125 we have daughter plates of the two classes of second spermatocytes, showing the content of the two equal classes of dimorphic spermatozoa, as this was shown in *Tenebrio*. Figures 126 and 127 are anaphases showing the division of the heterochromosomes (*l* and *s*). Figures 128-130 are early stages in the development of the spermatid showing the chromatin nucleolus (*n*) in various phases.

Blepharida rhois (Family Chrysomelidæ).

The testes were rather too far advanced when this material was collected and no dividing spermatogonia were present. The growth stages (figs. 131, 132) show a faintly staining spireme and a heterochromosome group similar to that of *Odontota*, a large and a small chromosome attached to a large plasmosome. The spireme appears to go directly over by condensation and segmentation into the dumb-bell-shaped figures seen in the first maturation spindle (figs. 133, 134), though cross-shaped bivalents occasionally occur (fig. 135). The heterochromosome pair, slightly separated by plasmosome material, is usually found at the periphery of the plate (figs. 133-136). Figure 137 is an exceptional anaphase in which the heterochromosome elements are not mingled with the polar masses of chromatin. Figures 138 *a* and *b* are equatorial plates of the second mitosis, and figures 139 and 140 are pairs of daughter plates from second spermatocytes showing again the dimorphism of the spermatozoa as to their chromatin content. As in several of the forms studied, material was collected for examination of the somatic cells, but no favorable cases of mitosis were to be found.

Silpha americana (Family Silphidæ).

Only one male of this species was secured, but the large testes gave all stages in abundance. The chromosomes, however, were very small and too numerous, 40 in the spermatogonia (fig. 141). The small chromosome is, nevertheless clearly distinguished in many of these plates (*s*). The resting spermatogonium contains one very large plasmosome and often one or two smaller ones (fig. 142, *p*). The unequal pair is seen in the growth stages (figs. 143, 144), and may frequently be seen outside of the equatorial plate of the first spermatocyte spindle (fig. 146). In favorable sections it may also be found in the plate among the other bivalents (fig. 147). Figure 145 is a prophase showing the bivalent chromosomes still connected by linin fibers. An equatorial plate of the first division is shown in figure 148, and a pair of corresponding plates of the second spermatocyte in figure 149. The small heterochromosome divides in the second spindle in advance of the others as seen in figure 150. Therefore, although this form is not especially favorable for detailed study on account of the large number of small chromosomes, the conditions are evidently the same as in the other species described—an unsymmetrical heterochromosome bivalent in the first spermatocyte, giving rise by the second maturation division to equal numbers of dimorphic spermatozoa, one class receiving the large heterochromosome, the other class the small one.

Doryphora decemlineata (Family Chrysomelidæ).

Doryphora decemlineata has been the most difficult one of the collection to work out satisfactorily. The chromosomes in the spermatogonial plates were in most cases much tangled, and the behavior of the heterochromosome pair was such as to suggest an "accessory chromosome" rather than an unequal pair. Abundant material for the study of somatic cells was at hand, but nothing favorable could be found in the sections.

Two spermatogonial plates, containing 36 chromosomes, are shown in figures 151 and 152 (plate XII). The small heterochromosome (*s*) is slightly elongated. The synizesis and synapsis stages are especially clear. The chromosomes, after the last spermatogonial mitosis go over immediately into a synizesis stage consisting of a polarized group of short loops, which later straighten and unite in pairs (figs. 153 and 154). From these loops are formed the spireme (figs. 155-158), which splits and segments, producing various cross, dumb-bell, and ring forms (figs. 159-163). As in most of the other species of Coleoptera, the unequal pair is not distinguishable until the spireme stage. Figure 162 is an unusual prophase in which all of the equal pairs show a longitudinal split as well as a transverse constriction, and the larger heterochromosome (*l*) is also split. Figure 163 shows a somewhat later and more common prophase in which the unequal pair, one ring, crosses, and dumb-bells may be seen. This figure, as well as figures 164-168, show the unequal pair in various relations to the other chromosomes. This pair in *Doryphora* consists of a large V-shaped chromosome with a small spherical one attached to it in different positions. When the small one is behind the V, the group has the appearance of an orthopteran "accessory."

Figures 169-171 show the separation of the two elements outside of the equatorial plate, while in figure 168 the unequal pair is in line with the other chromosomes. In figure 172, an anaphase, the unequal elements are barely separated, while the metakinesis of the other pairs is much further advanced.

Figures 173 and 174 are equatorial plates of the first division, one showing only the larger element of the heterochromosome pair (fig. 174, *x*), the other both elements (fig. 173, *l* and *s*). In the late anaphase (fig. 175) the larger heterochromosome is often seen outside of the polar mass, reminding one again of the "accessory" in the Orthoptera. Occasionally it is found in some other isolated position (fig. 176). Equatorial plates of the second division show the same conditions as in the other species; some contain the larger heterochromosome, others the smaller one (fig. 177, *a* and *b*). It was impossible to draw anaphases of the second division from a polar view and the lateral view showed nothing unusual, merely the longitudinal division of all of the chromosomes.

The spermatids show some interesting variations from the other species which have been examined. In figures 178 and 179 we have telophases of the second spermatocyte, showing centrosome and archoplasm (fig. 178) and certain masses of deeply staining material in the cytoplasm (fig. 179, a_1). Figures 180 and 181 are young spermatids showing the archoplasm from the second spindle (a_2) and a smaller, more deeply staining mass (a_1), derived from the irregular masses of the earlier stage (fig. 179, a_1). In figures 182 and 183, the axial fiber has appeared and the larger mass of archoplasm (a_2) is being transformed into a sheath. The other body remains unchanged. During the following stages this smaller archoplasmic body (a_1) lies in close contact with the axial fiber and sheath (a_2), and gradually decreases in size (figs. 184-186) until it disappears in a slightly later stage. The acrosome seems to develop directly out of the cytoplasm. The enigmatical body (a_1), which is probably archoplasm from the first maturation spindle, as it is not found in the cytoplasm of the first spermatocyte, may serve as nutriment for the developing axial fiber. The sperm head has a peculiar triangular form, staining more deeply on two sides.

Miscellaneous Coleoptera.

Considerable material from the spruce borers was collected at Harpswell, Maine, but the species were not identified. Although these insects were in the pupa stage, most of the testes were too old. There were no dividing spermatogonia and few spermatocyte mitoses. Most of the spermatocytes contained 10 chromosomes, one of which was plainly an unequal pair. In a few testes the number was 11, indicating that pupæ of two species had been collected. Figure 187 shows the metaphase of first spermatocyte mitosis with the unequal pair in metakinesis. Figures 188 and 189 are first spermatocyte equatorial plates of the two species, containing 10 and 11 chromosomes respectively. Figure 190 is a first spermatocyte spindle in anaphase, showing the unequal pair behind the other chromosomes. Figure 191 is an equatorial plate from a second spermatocyte, showing the small chromosome. In figure 192 are shown several of the bivalent chromosomes, including the unsymmetrical pair, from nuclear prophases of the first division, all from the same cyst.

Adalia bipunctata (family Coccinellidæ), the common lady beetle, has a very conspicuous pair of unequal heterochromosomes, as may be seen in figures 193-197 (plate XIII). This would seem to be a favorable form for determining the chromosome conditions in somatic cells, but no clear equatorial plates were found in either larvæ or pupæ.

In *Cicindela primeriana* (family Cicindelidæ) there are 18 chromosomes in the spermatogonium (fig. 198), one being small. The heterochromosome group is blended into a vacuolated sphere in growth stages (figs. 199, 200). In the metaphase of the first division it is trilobed, or tripartite (fig. 201), and in metakinesis, a small spherical chromosome separates from a much larger V-shaped one (fig. 202). Equatorial plates of first and second spermatocytes are shown in figures 203 and 204. Whole cysts of giant first spermatocytes were found both in growth stages (fig. 205) and prophases (fig. 206). Here the heterochromosome group is plainly double (fig. 205), and the conditions observed must have been due to the failure of a spermatogonial mitosis to complete itself.

Several of the Carabidæ have been studied, and the material, though not especially favorable, is interesting in that some members of the family have an unequal pair of heterochromosomes, others an odd one. *Chlænius æstivus* (figs. 207-212), *Chlænius pennsylvanicus* (figs. 213-215), and *Galerita bicolor* (fig. 216) have the unequal pair, while *Anomoglossus emarginatus* (figs. 217-223) has an odd heterochromosome (x), which behaves exactly like the larger heterochromosome in other carabs.

In the Elateridæ and Lampyridæ we also have examples of the second type with the odd chromosome. Two Elaters, species not determined (figs. 224-

229 and 230-235), have each 19 chromosomes in the spermatogonia (figs. 224 and 230), and in the first spermatocyte division an odd chromosome (*x*) which is in each case the smallest. In the first of these Elaters, the female somatic number was determined to be 20 (fig. 229). In the second Elater the pairs of second spermatocytes, containing 9 and 10 chromosomes respectively in the two cells, were in nearly every case connected as shown in figure 235, one pair of chromosomes not having separated completely in the first mitosis. Of *Ellychnia corrusca* (family Lampyridæ) only the spermatogonial equatorial plate, containing 19 chromosomes (*x*, the odd one) is given, as no material in maturation has yet been obtained, and a comparative study of the germ cells of the Elateridæ and Lampyridæ will be made as soon as suitable material can be secured.

In addition to the species of Coleoptera described here, two others, *Coptocycla aurichalcea* and *Coptocycla guttata* have been studied by one of my students and the results published elsewhere (Nowlin, '06). In both an even number of chromosomes (22, 18) was found in the spermatogonia, one being very small and forming with a larger one an unequal pair which remained condensed during the growth stage and separated into its larger and smaller components in the first spermatocyte mitosis. The result of maturation, as in the other species here described and in *Tenebrio molitor*, is dimorphism of the spermatozoa. The method of synapsis in Coptocycla is like that described for *Chelymorpha argus*.

HEMIPTERA HOMOPTERA.

Aphrophora quadrangularis.

The abundance of Aphrophora at Harpswell, Maine, in June and July, 1905, suggested that it might be well to examine at least one more of the Hemiptera homoptera for comparison with the many species of Hemiptera heteroptera which have been recently reexamined by Wilson ('05, '05, '06).

The larvæ only were collected, as they gave all the desired stages for a study of the spermatogenesis, and also oögonia and synizesis and synapsis stages of the oöcytes. In the first collections the testes were dissected out, but the many free follicles break apart so easily that the later material was prepared by cutting out the abdominal segments which contained the reproductive organs, and fixing those without dissection. The same methods of fixation and staining were employed as for the Coleoptera. Hermann's safranin-gentian method was especially effective with this material.

In *Aphrophora* the follicles of each testis are free, forming a dense cluster, each follicle being connected with the vas deferens by a short duct. The very young follicles are spherical, the older ones ovoid in form. The primary spermatogonia (plate XIV, fig. 237)—very clear cells with a lobed nucleus which stains slightly—occupy the tip of the follicle. Next to these comes a layer of cysts of secondary spermatogonia which are conspicuous for their deeper staining quality (fig. 238). There appears to be no plasmosome in either class of spermatogonia. Figure 239 is the equatorial plate of a secondary spermatogonium. There are 23 chromosomes, two of which are conspicuously larger than the others and evidently form a pair. The odd one is one of the three next in size.

Next to the secondary spermatogonia are cysts of young spermatocytes, whose nuclei show a continuous spireme and an elongated deeply staining chromatin rod which is the odd chromosome (fig. 240). This is often more elongated than in the figure and more or less wormlike in appearance. A pair of smaller chromatin masses may sometimes be detected at this stage, and are readily found a little later (fig. 241) when the nucleus has enlarged and the spireme has become looser and stains less deeply. Here the odd chromosome is more condensed, or shortened, and split. There is no synizesis and no polarized or bouquet stage, but the nuclei of all of the spermatocytes contain a continuous spireme throughout the growth stage. Synapsis must occur at the close of the last spermatogonial mitosis before the spireme is formed. Figures 242 and 243 show a slightly later growth stage. The form and connection of the "*m*-chromosome" pair (Wilson, '05$_b$)

comes out clearly here. Figure 244, from a safranin-gentian preparation, shows both the odd chromosome and the *m*-chromosomes. Some time before the first mitosis, the spireme splits and the pairs of granules embedded in linin are wonderfully distinct, both in iron-hæmatoxylin and safranin-gentian preparations (fig. 245). The *m*-chromosomes have here formed a precocious tetrad (*m*). Figure 246 is a similar stage from a safranin-gentian preparation. Figures 247 and 248 show the condensation of chromatin granules to form tetrads of various sizes, still embedded in the linin spireme. As these tetrads come into the spindle without losing their elongated form, it is evident that each one consists of two longitudinally split chromosomes united end to end in synapsis and separated in the first maturation mitosis, which is therefore reductional. The odd chromosome and the *m*-chromosomes show no longitudinal split in these figures, but they may appear as in figure 249. Occasionally one of the tetrads takes the form of a cross (fig. 249). In this figure the split "accessory" (*x*) lies against the nuclear membrane and the archoplasmic material for the spindle is seen along one side of the nucleus. It is certain here that the spindle fibers come from extranuclear material, not from nuclear substance, as Paulmier ('99) describes for *Anasa tristis*.

Figures 250 and 251 show the first maturation mitosis as it usually appears in sections from mercuro-nitric material stained with iron-hæmatoxylin. The odd chromosome is always more or less eccentric and is attached by a spindle fiber to one pole. In Hermann material, considerably destained, the tetrads and the odd chromosome appear as in figures 252, 253, and 254, the tetrads being in position for a transverse division. The odd chromosome is always so placed that its longitudinal split is at right angles to the axis of the spindle, as though it were to divide in this mitosis. It does not do so, however, but goes to one daughter cell, always lagging behind, as is shown in figures 255 and 256. Figures 257, *a* and *b*, are polar plates of the first mitosis with 11 and 12 chromosomes, respectively, and figures 258, *a*, *b*, and *c*, show the polar plates (*a* and *c*) each containing 11 chromosomes, and the odd chromosome at a different level (*b*). The latter is a view of the anaphase which one often gets at three foci in one section. Figures 259, *a* and *b*, are equatorial plates of the second mitosis with 11 and 12 chromosomes respectively. Figure 260 shows a side view of the second spindle in metaphase, and figure 261 in anaphase. Figures 262 and 263 are daughter plates from two spindles showing the chromosome content of the two equal classes of spermatozoa, one class containing 11 ordinary chromosomes, the other 11 ordinary chromosomes plus the odd heterochromosome, for the odd chromosome divides with the others in the second spindle as in Orthoptera (McClung and Sutton).

In figures 264 and 265 (plate XV) are seen the telophase of the two kinds of second spermatocytes, one (fig. 265) showing the divided odd chromosome, which continues to stain more deeply after the others have become diffuse. All of the spermatids (figs. 266-268) contain, in the early stages of development, a body (*n*) which stains like chromatin, but increases in size from a small granule in the telophase (figs. 264, 265) to the large dense body (n) seen in figure 267. This is probably homologous with the chromatin nucleolus described for the spermatids of the Coleoptera. In addition to this, in one-half of the spermatid nuclei there is a condensed mass of chromatin which is evidently the derivative of the odd chromosome of the spermatogonia and spermatocytes (figs. 267 and 268, x). In common with the spermatids of other Hemiptera these show two masses of archoplasm, the larger of which forms the sheath (s) of the axial fiber of the tail, and the smaller the acrosome (a). The axial fiber grows out directly from the centrosome, on either side of which there is a dense band forming the lateral boundary of the middle piece. It will be seen that the odd chromosome of Aphrophora is in its behavior precisely like the typical Orthopteran "accessory" of McClung, and similar to the odd chromosome of the Coleoptera.

In various parts of the young male larvæ dividing cells were found and the number 23 determined (fig. 269). Turning now to the female larvæ to determine the somatic number, the oögonia proved to be more favorable for counting. Twenty-four chromosomes were present in equatorial plates of oögonial mitoses (fig. 270), thus confirming Wilson's results for the *Anasa* group of the Hemiptera heteroptera.

In examining sections of female larvæ stained with safranin-gentian-violet, I was surprised to see a very marked polarized or bouquet stage and to find among the loops something resembling the odd chromosome of the growing spermatocytes. It was difficult to get a clear view of this body as it lay within the loops. In one section of a slightly earlier stage before synapsis, there were found two pairs of chromosomes (fig. 271, x_1, x_2, and m_1, m_2) which were stained with safranin in contrast with the violet spireme. These two pairs I interpret as being (1) the homologues of the pair of m-chromosomes, which remain condensed during the growth stage of the spermatocytes, and (2) a pair of heterochromosomes corresponding to the odd chromosome of the male. Various combinations of these heterochromosomes are shown in figures 272-277. Figures 278 and 279 were taken from mercuro-nitric material stained with iron-hæmatoxylin. In section 278 the "bouquet" was cut through, showing the bivalent corresponding to the larger pair in figure 271, and in figure 279 this element is seen behind the paler loops. The history of these two pairs of heterochromosomes, which have not, so far as I know, been found before in oöcytes, should be followed up in older ovaries, and related species should be examined for similar phenomena.

LEPIDOPTERA.

Cacoecia and Euvanessa.

I had no intention of making an extended study of the spermatogenesis of the Lepidoptera, but was interested to see if anything corresponding to the heterochromosomes of other orders could be found. The material studied was the testes of the larvæ of *Cacœcia cerasivorana* and *Euvanessa antiopa*. The number of chromosomes is large, but the equatorial plates are diagrammatically clear. In both species 30 chromosomes are found in both first and second spermatocytes. In both, one chromosome is larger (figs. 290 and 293, x). In the growth stage (figs. 283, 284) there is a two-lobed body (or sometimes two separate spherical bodies) which seems to correspond in size to the larger pair of chromosomes in the first spermatocyte. In iron-hæmatoxylin preparations this pair is often obscured by parts of the spireme which are tangled around it. In safranin-gentian preparations it stains, not like a plasmosome, but red like the heterochromosomes, while the spireme is violet. The staining reaction at least suggests that this equal pair of chromosomes, which may be traced through the synizesis stage (fig. 280), synapsis stage (figs. 281, 282), growth stages (figs. 283, 284), and prophases (figs. 285-287), into the first spermatocyte spindle (figs. 288, 290), and on to the second spermatocyte (figs. 289, 291, 292), is an equal pair of heterochromosomes comparable to the equal pair of "idiochromosomes" found by Wilson in *Nezara* ('05). As the various stages are practically the same in *Euvanessa antiopa*, but somewhat clearer in *Cacœcia*, only one figure is given for *Euvanessa*—the equatorial plate of the first spermatocyte (fig. 293).

SUMMARY OF RESULTS.

(1) An unequal pair of heterochromosomes has been found by the author in 19 species of Coleoptera belonging to 8 families:

	Family.	Species.
I.	Buprestidæ	Two spruce-borers, species not determined.
II.	Carabidæ	{ 1. *Chlænius æstivus.* { 2. *Chlænius pennsylvanicus.* { 3. *Galerita bicolor.*
III.	Chrysomelidæ	{ 1. *Blepharida rhois.* { 2. *Chelymorpha argus.* { 3. *Coptocycla aurichalcea.* { 4. *Coptocycla guttata.* { 5. *Doryphora decemlineata.* { 6. *Odontota dorsalis.* { 7. *Trirhabda virgata.* { 8. *Trirhabda canadense.*

IV. Cicindelidæ   *Cicindela primeriana.*

V.   Coccinellidæ   { *Adalia bipunctata.*
                     { *Epilachna borealis.*

VI. Scarabæidæ   *Euphoria inda.*

VII. Silphidæ   *Silpha americana.*

VIII. Tenebrionidæ   *Tenebrio molitor.*

(2) An odd chromosome, which behaves during the growth stage of the first spermatocytes like the "accessory" of the Orthoptera, has been found in 4 species of Coleoptera,[A] belonging to 3 families:

Family.   Species.

I. Carabidæ   *Anomoglossus emarginatus.*

II. Elateridæ   Two Elaters; species not determined.

III. Lampyridæ   *Ellychnia corrusca.*

(3) In most of the species of Coleoptera examined, the unequal pair or the odd chromosome remains condensed during the growth period of the first spermatocyte, like the "accessory" of the Orthoptera and the various heterochromosomes of the Hemiptera.

(4) Several of these species of Coleoptera have a synizesis stage in which the spermatogonial number of short loops is massed at one side of the

nucleus. This is followed by a synapsis stage in which the loops straighten and unite in pairs, forming longer loops which soon spread out in the nuclear space, and, with the exception of the heterochromosomes, unite to form a continuous spireme.

(5) In several of the species of Coleoptera and in Aphrophora, it has been shown that a body staining like chromatin develops in the spermatids, increasing in size for a time, then breaking up into granules and disappearing. This body evidently has no relation to the heterochromosomes, as it is the same for all of the spermatids. Its staining qualities suggest that it may be material derived from the chromosomes. It is finally dissolved in the karyolymph.

(6) In iron-hæmatoxylin preparations the heterochromosomes of the Coleoptera vary greatly in their staining properties during mitosis. In some species they stain exactly like the ordinary chromosomes, in others the larger one of the unequal pair holds the stain more tenaciously than the others and also than its smaller mate, and this is true in several cases where the heterochromosome is smaller than the other chromosomes, which destain more readily. The odd chromosome of the Elaters stains less deeply than the others in the first spermatocyte. In the growth stage they stain more deeply, as a rule, than the spireme, with iron-hæmatoxylin or thionin, stain red with safranin-gentian and green with Auerbach's methyl green-fuchsin combination.

(7) *Aphrophora quadrangularis* agrees with the *Anasa* group of Hemiptera heteroptera in having a pair of *m*-chromosomes and an odd chromosome in the spermatocytes, but differs from many of that group in that the odd chromosome divides in the second mitosis instead of the first. It also differs from other known forms in exhibiting heterochromosomes in certain stages of the oöcytes.

(8) The two species of Lepidoptera examined have an equal pair of heterochromosomes.

FOOTNOTES:

[A] AUG. 20, 1906.—Since this paper was prepared, 19 other species of Coleoptera have been studied. Of these, 17 have an unequal pair of heterochromosomes in the spermatocytes. Six belong to the Chrysomelidæ, making 14 of that family that have been examined. Representatives of 4 new families—Melandryidæ, Lamiinæ, Meloidæ, Cerambycinæ have been studied. In only two species—1 Elater and 1 Lampyrid—has the odd chromosome been found in place of the unequal pair. No species of Coleoptera has yet been examined in which one or the other of these two types of heterochromosomes does not occur in the spermatocytes. Of the

42 species of Coleoptera whose germ cells have been studied, 85.7 per cent are characterized by the presence of an unequal pair of heterochromosomes in the male germ cells, 14.3 per cent by the presence of an odd chromosome.

COMPARISON OF RESULTS IN DIFFERENT SPECIES OF COLEOPTERA.

In number of chromosomes there is great variation, the smallest number (16) having been found in *Odontota dorsalis*, and the largest (40) in *Silpha americana*. The difference in size is also very marked, as may be seen by comparing the spermatogonial plates in figures 3 and 58 with those shown in figures 94 and 141.

No other species of the Tenebrionidæ has yet been secured, and all of the other beetles examined differ in a marked degree from *Tenebrio molitor* in the growth stages of the spermatocytes. While in *Tenebrio* the chromatin stains very dark throughout the growth stage, and the unequal pair can not be distinguished until the prophase of division ('05, plate VI, figs. 171-180), in most of the others there are very distinct synizesis and synapsis stages, following the last spermatogonial mitosis, then a spireme stage in which the condensed unequal pair of heterochromosomes or the odd chromosome is conspicuous in contrast with the pale spireme, whether the preparation is stained with iron-hæmatoxylin, gentian, or thionin. In *Tenebrio molitor*, the unequal pair behaved in every respect like the other bivalent chromosomes. In the other forms, though it behaves during the two maturation divisions like the symmetrical bivalents, it remains condensed during the growth period like the "accessory" of the Orthoptera, the odd chromosome, "*m*-chromosomes," and "idiochromosomes" of the Hemiptera. In several cases the heterochromosomes of the Coleoptera are associated with a plasmosome (figs. 22, 23, 63, 132, 158, 217), as is often true in other orders. This peculiar pair of unequal heterochromosomes varies considerably in size during the growth stage in some of the species studied, but changes very little in form, differing in this respect from the "accessory" in some of the Orthoptera (McClung, '02) and from the large idiochromosome in some of the Hemiptera (Wilson, '05).

The odd chromosome, so far as it has been studied, behaves precisely like the larger member of the unequal pair without its smaller mate (figs. 219, 220, 226, 233). In the growth stage it remains condensed and either spherical or sometimes flattened against the nuclear membrane (figs. 217, 225, 231). In the first maturation mitosis it is attached to one pole of the spindle, does not divide, but goes to one of the two second spermatocytes (figs. 233, 235). In the second spermatocyte it divides with the other chromosomes, giving two equal classes of spermatids differing by the presence or absence of this odd chromosome.

All of the evidence at hand leads to the conclusion that in the Coleoptera, the univalent elements of all the pairs, equal and unequal, separate in the first spermatocyte mitosis and divide quantitatively in the second. In this respect the behavior of the chromosomes in this order appears to be much more uniform than in the Orthoptera and Hemiptera.

COMPARISON OF THE COLEOPTERA WITH THE HEMIPTERA AND LEPIDOPTERA.

As has been seen above, the conditions in the Coleoptera, so far as the heterochromosomes are concerned, correspond very closely in final results with those in the Hemiptera heteroptera and the Orthoptera. In minor details these chromosomes are less peculiar in the Coleoptera than in either of the other orders. Even condensation during the growth stage is not universal, and synapsis of the heterochromosomes apparently occurs simultaneously with that of the ordinary chromosomes, instead of being delayed, as in many of the Hemiptera heteroptera.

Aphrophora (Hemiptera homoptera) agrees with the *Anasa* group of the Hemiptera heteroptera in having a pair of condensed m-chromosomes, in the growth stage, but this pair is already united in synapsis when first seen. It differs from *Anasa*, but agrees with *Banasa* and *Archimerus* in exhibiting a typical odd chromosome which goes to one pole without division in the first spermatocyte, and divides with the other chromosomes in the second spermatocyte. The odd chromosome in this species of Hemiptera, therefore, behaves like that in the Coleoptera and Orthoptera. The most interesting points in the results of this study of the germ cells of *Aphrophora* is the discovery of two pairs of condensed chromosomes in certain phases of the growth stages of the oöcytes. This has not been shown to be the case in any other species of Hemiptera, so far as I can ascertain. It is now evident that in the Heteroptera homoptera there are at least two distinct classes as to behavior of chromosomes. In one class we have the Aphids (Stevens, '05 and '06) and Phylloxera (Morgan, '06) in which no heterochromosomes have been found, while in the other class are such forms as Aphrophora with both a pair of m-chromosomes and a typical odd heterochromosome.

The two species of Lepidoptera examined indicate that here we may have conditions comparable to those in *Nezara*—an equal pair of heterochromosomes whose only apparent peculiarity is their condensed form during the growth stage. Doubtless the results of other investigators will soon throw more light on the heterochromosomes of this order.

GENERAL DISCUSSION.

It will be seen from the foregoing that the results obtained in the study of the germ cells of *Tenebrio molitor* have been confirmed in full for several species of Coleoptera, and in part for 19[B] different species belonging to 8[B] families. It has also been shown that a different type of Coleopteran spermatogenesis exists in at least 3 families, where an odd chromosome like that in the Orthoptera occurs in place of the unequal pair. In all of these insects the spermatozoa are distinctly dimorphic, forming two equal classes, one of which either contains one smaller chromosome or lacks one chromosome.

The most difficult part of the work has been the determination of the somatic number of chromosomes in the male and female. In some cases suitable material has been lacking; in others, though material was abundant, no metaphases could be found in which the chromosomes were sufficiently separated to be counted with certainty. In three species (in addition to *Tenebrio molitor*) where the unequal pair is present, the female somatic cells have been shown to contain the same number of chromosomes as the spermatogonia, but an equal pair in place of the unequal pair of the male. In two new cases the male somatic number and size have been shown to be the same as in the spermatogonia. In one of the Elateridæ, where the spermatogonial number is 19, the female somatic number is 20, and in *Aphrophora* the numbers in male and female cells are respectively 23 and 24. No exception has been found to the rule established by previous work on the Coleoptera (Stevens, '05) and on the Hemiptera (Wilson, '05 and '06), that (1) in cases where an unequal pair is present in the male germ cells, it is also present in the male somatic cells, but is replaced in the female by an equal pair, each component being equal in volume to the larger member of the unequal pair, and (2) in cases where an odd chromosome occurs in the male, a pair of equal size are found in the female. It is therefore evident that an egg fertilized by a spermatozoön (1) containing the small member of an unequal pair or (2) lacking one chromosome, must develop into a male, while an egg fertilized by a spermatozoön containing the larger element of an unequal pair of heterochromosomes or the odd chromosome must produce a female.

Whether these heterochromosomes are to be regarded as sex chromosomes in the sense that they both represent sex characters and determine sex, one can not decide without further evidence.

Comparison of the two types in Coleoptera, especially where, as in the Carabidæ, both occur in one family, has suggested to me that here it is possible that the small chromosome represents not a degenerate female sex chromosome, as suggested by Wilson, but some character or characters which are correlated with the sex character in some species and not in others. Assuming this to be the case, a pair of small chromosomes might be subtracted from the unequal pair, leaving an odd chromosome. The two types would then be reduced to one. It may be possible to determine the validity of this suggestion for particular cases by observation or experiment.

Since the first of this series of papers was published, there have appeared three important papers by Prof. E. B. Wilson, bearing on the problem of sex determination in insects. These papers are based on a study of many species of the Hemiptera heteroptera. These insects fall into two classes—one in which a pair of "idiochromosomes," usually of different size, remain separate and divide quantitatively in the first spermatocyte, conjugate and then separate in the second maturation mitosis; and another class in which an odd chromosome—the "heterotropic" chromosome—divides in one of the maturation mitoses, but not in the other. Wilson regards the odd chromosome as the equivalent of the larger of the "idiochromosomes," its smaller mate having disappeared. In the somatic cells of the former class he finds in the male the unequal pair, in the female an equal pair, the smaller chromosome being replaced by an equivalent of the larger "idiochromosome." In the latter class the male somatic cells contain the odd number, the female somatic cells and oögonia an even number, the homologue of the odd chromosome of the male being present and giving to the female one more chromosome than are found in the male.

In his latest paper Wilson ('06) makes a variety of suggestions as to sex determination. He shows that if the "idiochromosomes" and the heterotropic chromosome be regarded as sex chromosomes in the double sense that they both bear sex characters and determine sex, the following scheme accounts for the observed facts in all cases where an unequal pair or an odd heterochromosome have been found:

Sperm.   Egg

I.   {Large ♂ "idiochromosome" }
     {or   } +Large ♀ sex chromosome   =   a ♀
     {Odd chromosome.   }

II. {Small ♀ "idiochromosome" }

   {or   } + Large ♂ sex chromosome = a ♂

   {No sex chromosome   }

Here we know that such a combination of gametes must occur to give the observed results, but we are not certain that we have a right to attribute the sex characters to these particular chromosomes or in fact to any chromosomes. It seems, however, a reasonable assumption in accordance with the observed conditions. The scheme also assumes either selective fertilization or, what amounts to the same thing, infertility of gametic unions where like sex chromosomes are present. It also assumes that the large female sex chromosome is dominant in the presence of the male sex chromosome, and that the male sex chromosome is dominant in the presence of the small female sex chromosome. Or, it might rather be said that these are not really assumptions, but inferences as to what must be true if the heterochromosomes are sex chromosomes. This theory of sex determination brings the facts observed in regard to the heterochromosomes under Castle's modification of Mendel's Law of Heredity ('99).

The question of dominance is a difficult one, especially in parthenogenetic eggs and eggs which are distinctly male or female before fertilization. It may be possible that the sex character of the egg after maturation is always dominant in the fertilized egg, as appears to be the case in these insects (see scheme). Conditions external to the chromosomes may determine in certain cases, such as Dinophilus, which sex character shall dominate in the growing oöcyte, and maturation occur accordingly. It is evident that this reasoning would lead to the conclusion that sex is or may be determined in the egg before fertilization, and that selective fertilization, or infertility of gametic unions containing like sex characters, has to do, not with actual sex determination, but with suitable distribution of the sex characters to future generations. If both sex characters are present in parthenogenetic eggs, as appears to be the case in aphids and phylloxera, dominance of one or the other must be determined by conditions external to the chromosomes, for we have both sexes at different points in the same line of descent without either reduction or fertilization.

Wilson suggests as alternatives to the chromosome sex determinant theory according to Mendel's Law, (1) that the heterochromosomes may merely transmit sex characters, sex being determined by protoplasmic conditions external to the chromosomes; (2) That the heterochromosomes may be sex-determining factors only by virtue of difference in activity or amount of chromatin, the female sex chromosome in the male being less active. The first of these alternatives is an attempt to cover such cases as *Dinophilus*, *Hydatina*, and *Phylloxera* with large female and small male eggs. Here Morgan's ('06) suggestion as to degenerate males seems much to the point. The male sex character, having become dominant in certain eggs at an early stage, may, from that time on, determine the kind of development. As to the second alternative, I see no reason for supposing that the small heterochromosome of a pair is in any different condition, as to activity, from the large one. The condensed condition may not mean inactivity, but some special form of activity. And, moreover, it has been shown that in certain stages of the development of the oöcyte of one form, *Aphrophora quadrangularis*, there are pairs of condensed chromosomes corresponding to those of the spermatocyte, so that there would hardly seem to be any basis for Wilson's attempt to associate the difference in development of male and female germ cells with activity or inactivity of chromosomes, as indicated by condensed or diffuse condition of the chromatin.

On the whole, the first theory, which brings the sex determination question under Mendel's Law in a modified form, seems most in accordance with the facts, and makes one hopeful that in the near future it may be possible to formulate a general theory of sex determination.

This work has been done in connection with a study of the problem of sex determination, but, whatever may be the final decision on that question, it brings together a mass of evidence in favor of the belief in both morphological and physiological individuality of the chromosomes, as advocated by Boveri, Sutton, and Montgomery. It also gives the strongest kind of evidence that maternal and paternal homologues unite in synapsis and separate in maturation, leaving the ripe germ cells pure with regard to each pair of characters.

BRYN MAWR COLLEGE, *June 7, 1906.*

FOOTNOTES:

[B] AUG. 20, 1906.—36 species belonging to 12 families. See note, p. 49.

BIBLIOGRAPHY.

BOVERI, TH.

'02. Ueber mehrpolige Mitosen als Mittel zur Analyse des Zellkerns. Verh. d. phys.-med. Ges. Würzburg, N. F., vol. 35.

CASTLE, W. E.

'03. The heredity of sex. Bull. Mus. Comp. Zoöl. Harvard College, vol. 40, no. 4.

MCCLUNG, C. E.

'99. A peculiar nuclear element in the male reproductive cells of insects. Zoöl. Bull., vol. 2.

'00. The spermatocyte divisions of the Acridiidæ. Kans. Univ. Quart., vol. 9, no. 1.

'01. Notes on the accessory chromosome. Anat. Anz., vol. 20, nos. 8 and 9.

'02. The accessory chromosome—sex-determinant? Biol. Bull., vol. 3, nos. 1 and 2.

'02*a*. The spermatocyte divisions of the Locustidæ. Kans. Univ. Quart., vol. 1, no. 8.

'05. The chromosome complex of orthopteran spermatocytes. Biol. Bull., vol. 9, no. 5.

MONTGOMERY, THOS. H., JR.

'01. A study of the chromosomes of the germ-cells of Metazoa. Trans. Amer. Phil. Soc., vol. 20.

'03. The heterotypic maturation mitosis in Amphibia and its general significance. Biol. Bull., vol. 4, no. 5.

'01*a*. Further studies on the chromosomes of the Hemiptera heteroptera. Proc. Acad. Nat. Sci. Phila., 1901.

'04. Some observations and considerations upon the maturation phenomena of the germ-cells. Biol. Bull., vol. 6, no. 3.

'05. The spermatogenesis of *Syrbula* and *Lycosæ* and general considerations upon chromosome reduction and heterochromosomes. Proc. Acad. Nat. Sci. Phila., 1905.

MORGAN, T. H.

'06. The male and female eggs of Phylloxerans of the Hickories. Biol. Bull., vol. 10, no. 5.

NOWLIN, W. N.

'06. A study of the spermatogenesis of *Coptocycla aurichalcea* and *Coptocycla guttata*. Journ. of Exp. Zoöl., vol. 3, no. 3.

PAULMIER, F. C.

'99. The spermatogenesis of *Anasa tristis*. Journ. of Morph., vol. 15.

DE SINÉTY.

'01. Recherches sur la biologie et l'anatomie des phasms. La Cellule, vol. 19.

STEVENS, N. M.

'05. A study of the germ cells of *Aphis rosæ* and *Aphis œnotheræ*. Journ. of Exp. Zoöl., vol. 2, no. 3.

'05*a*. Studies in spermatogenesis, with especial reference to the "accessory chromosome." Carnegie Inst. of Wash., pub. no. 36.

'06. Studies on the germ cell of Aphids. Ibid., pub. no. 51.

SUTTON, W. S.

'02. On the morphology of the chromosome group in *Brachystola magna*. Biol. Bull., vol. 4, no. 1.

'03. The chromosomes in heredity. Biol. Bull., vol. 4, no. 5.

WILSON, E. B.

'05. Studies on chromosomes. I. The behavior of the idiochromosomes in Hemiptera. Journ. Exp. Zoöl., vol. 2, no. 3.

'05*a*. The chromosomes in relation to the determination of sex in insects. Science, vol. 22, no. 564.

'05*b*. Studies on chromosomes. II. The paired microchromosomes, idiochromosomes, and heterotropic chromosomes in Hemiptera. Journ. Exp. Zoöl., vol. 2, no. 4.

'06. Studies on chromosomes. III. The sexual differences of the chromosome-groups in Hemiptera, with some considerations of the determination and inheritance of sex. Ibid., vol. 3, no. 1.

DESCRIPTION OF PLATES

[The figures were all drawn with Zeiss oil-immersion 2 mm., oc. 12, and have been reduced one-third, giving a magnification of 1,000 diameters.]

PLATE VIII.

Trirhabda virgata (Family Chrysomelidæ).

FIG. 1. Equatorial plate from somatic tissues of a male pupa, 27 large chromosomes, 1 small one.

2. Equatorial plate from an egg follicle, 28 large chromosomes.

3. Equatorial plate of spermatogonium, 27 large chromosomes, 1 small one.

4. First spermatocyte, synizesis stage.

5. First spermatocyte, early spireme stage, showing unequal pair of chromosomes.

6-7. First spermatocyte, later growth stages.

8. First spermatocyte, prophase.

9-12. First spermatocyte, metaphase.

13. First spermatocyte, equatorial plate.

14-15. First spermatocyte, anaphase, showing separation of the elements of the unequal pair (*l* and *s*).

16. First spermatocyte, daughter plates.

17. Second spermatocytes, equatorial plates.

18. Second spermatocytes, equatorial plates showing V-shaped chromosomes.

19. Second spermatocyte, early anaphase, the small chromosome in metakinesis.

Trirhabda canadense.

20. Equatorial plate from egg follicle, 30 large chromosomes.

21. Equatorial plate of spermatogonium, 29 large chromosomes, 1 small one.

22. First spermatocyte, growth stage showing the heterochromosome group.

23. Heterochromosome group. p = plasmosome, l = large heterochromosome, s = small heterochromosome.

24-27. First spermatocyte, metaphase.

28. First spermatocyte, equatorial plate.

29. First spermatocyte, equatorial plate, small member of the unequal pair only present.

30. First spermatocyte, daughter plates.

31. Second spermatocytes, equatorial plates.

32-33. Second spermatocytes, prophase.

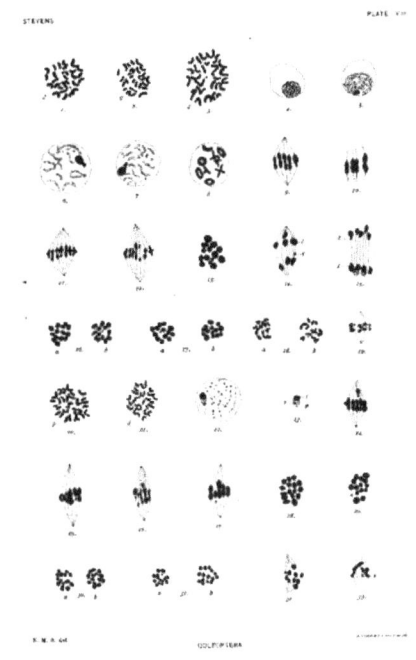

PLATE IX.

Chelymortha argus (Family Chrysomelidæ).

FIGS. 34-35. Equatorial plates from egg follicles, 11 equal pairs, no small chromosome.

36. Equatorial plate of spermatogonium, 21 large chromosomes, 1 small one.

37. First spermatocyte, synizesis stage.

- 31 -

38-40. First spermatocyte, synapsis stage.

41-43. First spermatocyte, bouquet stage after synapsis.

44-45. First spermatocyte, spireme stage showing the unequal pair of heterochromosomes.

46. First spermatocyte, prophase.

47-49. First spermatocyte, metaphase.

50-51. First spermatocyte, equatorial plates, x the heterochromosome pair.

52. First spermatocyte, showing metakinesis of the unequal pair.

53. First spermatocyte, anaphase.

54-55. Second spermatocyte, equatorial plates.

56. Second spermatocyte, anaphase.

Odontota dorsalis (*Family Chrysomelidæ*).

57. Equatorial plate of male somatic cell from walls of the testis, 15 large chromosomes, 1 small one.

58-59. Equatorial plates of spermatogonia, 15 large chromosomes, 1 small one.

60. Resting nucleus of spermatogonium, showing plasmosome (p).

61. First spermatocyte, synizesis stage.

62. First spermatocyte, synapsis stage.

63-64. First spermatocyte, spireme stage, showing the larger and smaller heterochromosome associated with a plasmosome.

65-68. First spermatocyte, prophases.

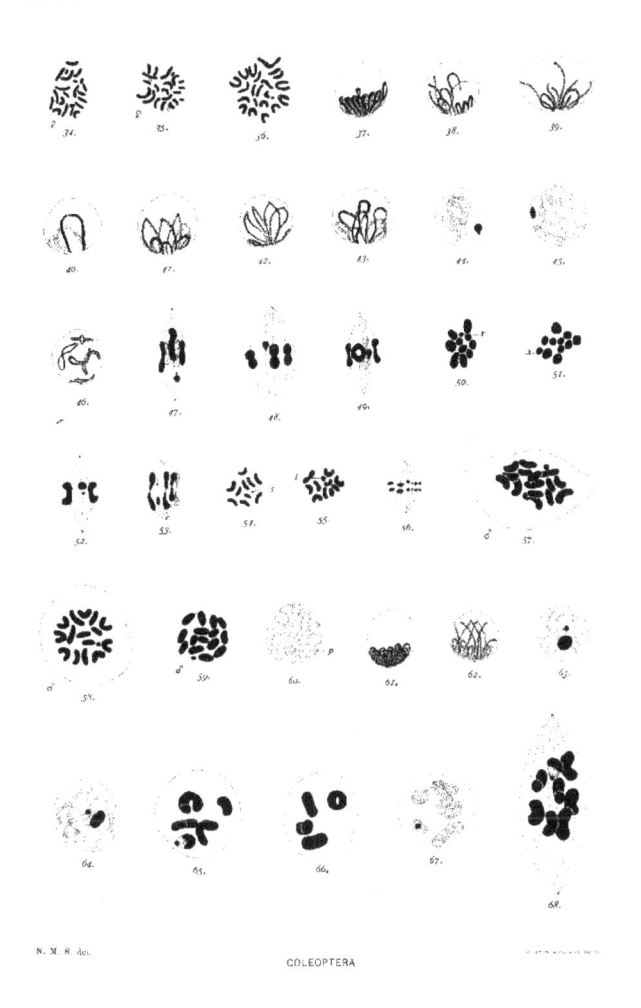

PLATE X.

Odontota dorsalis.

FIGS. 69-70. First spermatocyte, metaphase.

71. First spermatocyte, equatorial plate.

72. First spermatocyte, metaphase, showing metakinesis of the heterochromosomes.

73-74. First spermatocyte, anaphase.

75-76. Second spermatocyte, equatorial plates.

77. Second spermatocyte, showing metakinesis of the small chromosome (*s*).

78. Second spermatocyte, prophase, showing chromosomes longitudinally split.

79-80. Young spermatids, *n* the chromatin nucleolus.

81-87. A series of stages in the development of the sperm head, showing the various phases in the history of the chromatin nucleolus (*n*).

88. Cross-sections of nearly mature sperm heads.

89-90. Equatorial plates of spermatogonia of abnormal individual, 15 large chromosomes, 2 small ones.

91. First spermatocyte from same testis, spireme stage, showing 2 small chromosomes associated with 1 large one and a plasmosome.

92. First spermatocyte from the same testis, metaphase showing a similar heterochromosome group.

93. Second spermatocyte from same testis, equatorial plate, showing 2 small chromosomes.

Epilachna borealis (*Family Coccinellidæ*).

94. Equatorial plate of spermatogonium, 17 large chromosomes and 1 small one.

95. First spermatocyte, spireme stage, showing the unequal pair.

96-97. First spermatocyte, late prophases.

98. First spermatocyte, metaphase, showing chromosomes of different forms.

99-100. First spermatocyte, equatorial plate.

101. Unequal heterochromosome pair from a metaphase.

102. First spermatocyte, anaphase; ordinary chromosomes stippled to show more clearly the metakinesis of the unequal pair.

103. Second spermatocyte, equatorial plates.

104. Second spermatocyte, prophase.

105-106. Abnormal giant spermatids, probably in process of degeneration.

Euphoria inda (*Family Scarabæidæ*).

107. Equatorial plate of spermatogonium, 20 chromosomes. The 2 smallest are the unequal pair of heterochromosomes (*l* and *s*).

108. Resting spermatogonium, showing plasmosome (*p*).

109. First spermatocyte, spireme stage.

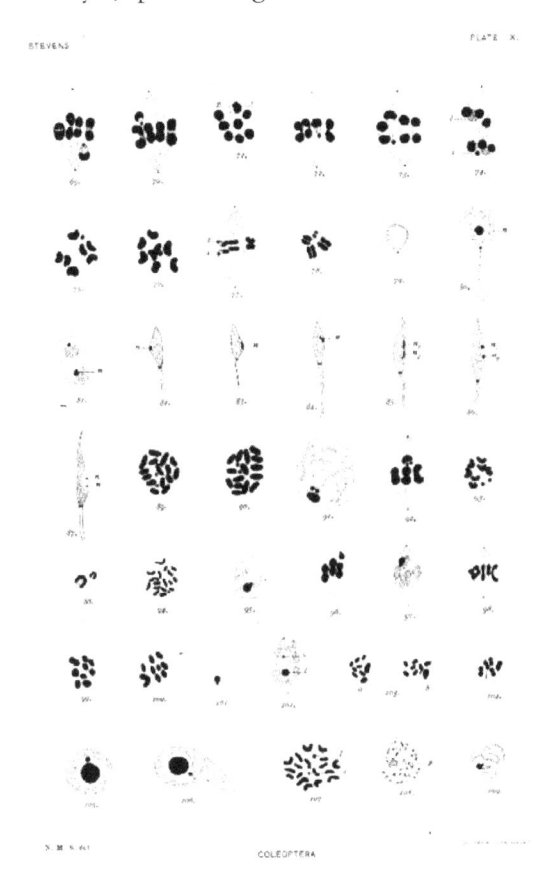

PLATE XI.

Euphoria inda.

FIGS. 110-111. First spermatocyte, prophases.

112-113. First spermatocyte, late prophase.

114-116. First spermatocyte, metaphase.

117. First spermatocyte, equatorial plate, *x* the unequal pair.

118-120. First spermatocyte, anaphase.

121. First spermatocyte, daughter plates.

122. Second spermatocyte, prophase.

123. Second spermatocyte, equatorial plates.

124-125. Second spermatocyte, daughter plates of the two classes.

126-127. Second spermatocyte, anaphase.

128-130. Spermatids, n the chromatin nucleolus.

Blepharida rhois (*Family Chrysomelidæ*).

131-132. First spermatocyte, spireme stages, showing the heterochromosome group.

133-135. First spermatocyte, beginning of metakinesis.

136. First spermatocyte, equatorial plate, x the unequal pair.

137. First spermatocyte, late anaphase, showing the heterochromosomes l and s.

138. Second spermatocyte, equatorial plates.

139-140. Second spermatocyte, daughter plates of the two classes.

Silpha americana (*Family Silphidæ*).

141. Equatorial plate of spermatogonium, 40 chromosomes—39 large, 1 small.

142. Resting nucleus of spermatogonium, showing 2 plasmosomes (p).

143-144. First spermatocyte, spireme stage.

145. First spermatocyte, prophase.

146-147. First spermatocyte, metaphase.

148. First spermatocyte, equatorial plate.

149. Second spermatocyte, equatorial plates.

150. Second spermatocyte, showing metakinesis of the small chromosome.

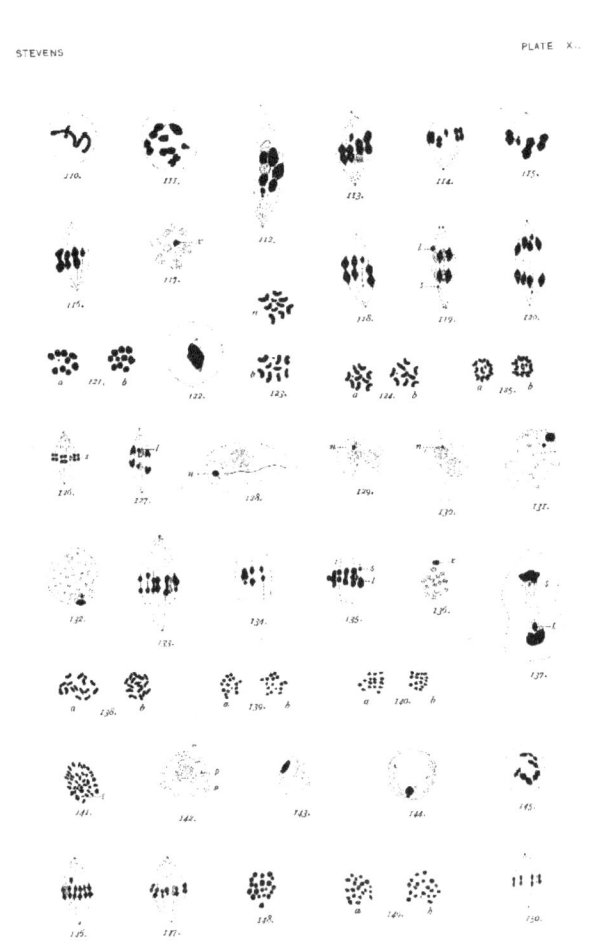

PLATE XII.

Doryphora decemlineata (*Family Chrysomelidæ*).

FIGS. 151-152. Equatorial plates of spermatogonia, 36 chromosomes—35 large, 1 small.

153. First spermatocyte, synizesis stage.

154. First spermatocyte, synapsis stage.

155-158. First spermatocyte, spireme stages.

159. First spermatocyte, spireme segmented and split.

160-163. First spermatocyte, prophases.

164-171. First spermatocyte, metaphase.

172. First spermatocyte, anaphase.

173-174. First spermatocyte, equatorial plates.

175-176. First spermatocyte, late anaphase.

177. Second spermatocyte, equatorial plates.

178-179. Second spermatocyte, telophase, a_1, archoplasmic material.

180-186. Spermatids in different stages; a_1, archoplasmic material from first spermatocyte spindle, a_2 archoplasmic material from second maturation spindle.

Spruce-borers (*Family Buprestidæ*).

187. First spermatocyte, metaphase.

188-189. First spermatocyte, equatorial plates of two species, 10 and 11 chromosomes.

190. First spermatocyte, anaphase.

191. Second spermatocyte, equatorial plates containing 9 large chromosomes and 1 small one.

192. Chromosomes from prophase of the first spermatocyte, all from the same cyst.

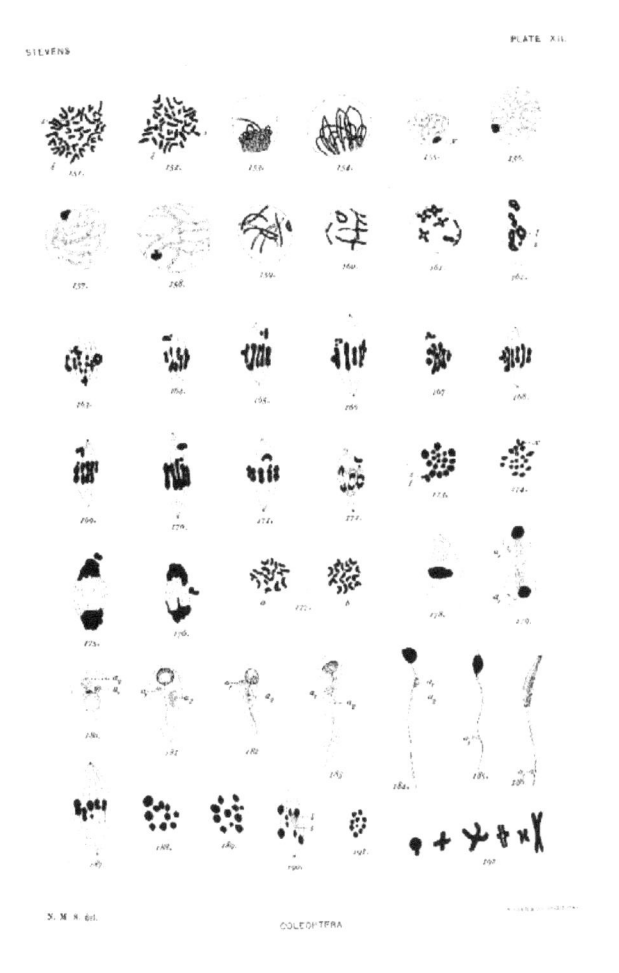

PLATE XIII.

Adalia bipunctata (Family Coccinellidæ).

FIG. 193. Equatorial plate of spermatogonium, 20 chromosomes—19 large, 1 small.

194. First spermatocyte, spireme stage, *x* the heterochromosome group.

195. First spermatocyte, metaphase.

196. First spermatocyte, equatorial plate.

197. Second spermatocyte, equatorial plates.

Cicindela primeriana (Family Cicindelidæ).

198. Equatorial plate of spermatogonium, 20 chromosomes—19 large, 1 small.

199. First spermatocyte, spireme stage, *x* the heterochromosome group.

200. First spermatocyte, prophase.

201. First spermatocyte, metaphase, *x* the unequal pair in tripartite form.

202. First spermatocyte, showing metakinesis of the heterochromosomes (*l* and *s*).

203. First spermatocyte, equatorial plate.

204. Second spermatocyte, equatorial plates.

205. Giant spermatocyte, spireme stage, heterochromosome group double the usual size.

206. Giant spermatocyte, prophase.

Chlænius æstivus (Family Carabidæ).

207. First spermatocyte, spireme stage, showing the unequal pair associated with a large plasmosome.

208. First spermatocyte, metaphase.

209-210. First spermatocyte, beginning of metakinesis.

211. First spermatocyte, equatorial plate, 17 chromosomes.

212. First spermatocyte, anaphase, showing elongated centrosome and diverging univalent chromosomes.

Chlænius pennsylvanicus.

213. First spermatocyte, spireme stage.

214. First spermatocyte, equatorial plate, *x* the unequal bivalent.

215. First spermatocyte, late prophase.

Galerita bicolor (Family Carabidæ).

216. Equatorial plate of spermatogonium, 30 chromosomes—29 large, 1 small.

Anomoglossus emarginatus (Family Carabidæ).

217. First spermatocyte, growth stage, *x* the odd chromosome.

218. First spermatocyte, prophase.

219-220. First spermatocytes, metaphase, *x* the odd chromosome.

221. First spermatocyte, equatorial plate.

222. First spermatocyte, daughter plates containing 18 and 19 chromosomes, respectively.

223. Second spermatocytes, equatorial plates.

Elater I (*Family Elateridæ, species not determined*).

224. Equatorial plate of spermatogonium, 19 chromosomes, x the odd one.

225. First spermatocyte, spireme stage, x the odd chromosome.

226. First spermatocyte, metaphase.

227. First spermatocyte, prophase.

228. First spermatocyte, equatorial plate.

229. Equatorial plate from egg follicle, 20 chromosomes, x_1 and x_2 the pair corresponding to x in the spermatogonium.

Elater II (*Species not determined*).

230. Equatorial plate of spermatogonium, 19 chromosomes, x the odd one.

231. First spermatocyte, spireme stage.

232. First spermatocyte, prophase.

233. First spermatocyte, beginning of metakinesis.

234. First spermatocyte, equatorial plate, x the odd chromosome.

235. A pair of second spermatocytes in metaphase, two chromosomes connected, x the odd chromosome.

Ellychnia corrusca (*Family Lampyridæ*).

236. Equatorial plate of spermatogonium, 19 chromosomes.

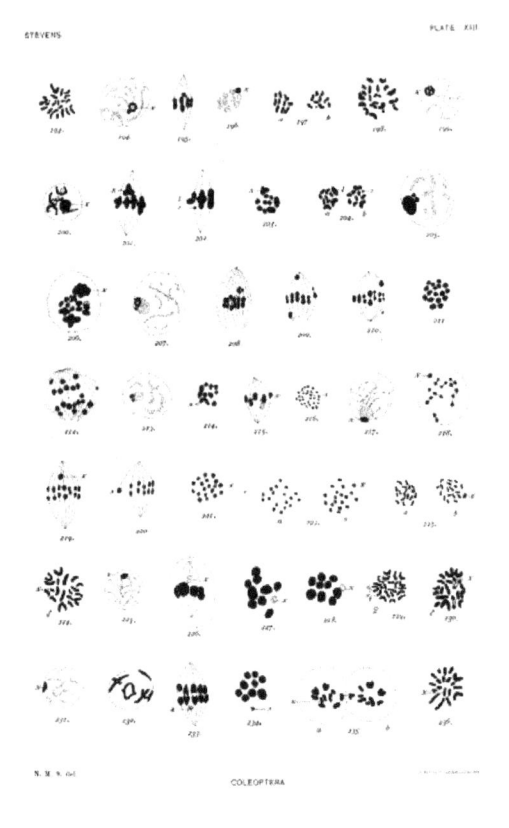

PLATE XIV.

Aphrophora quadrangularis (*Hemiptera homoptera*).

FIG. 237. Resting primary spermatogonium with lobed nucleus.

238. Resting secondary spermatogonium, with nucleus staining much more deeply.

239. Equatorial plate of secondary spermatogonium, 23 chromosomes.

240. First spermatocytes, very early growth stage, x the odd chromosome.

241-243. First spermatocyte, later spireme stages, showing the odd chromosome (x) and a pair of m-chromosomes (m).

244. Similar stage from a safranin-gentian preparation.

245. First spermatocyte, split-spireme stage, x the odd chromosome, m the m-chromosome tetrad.

246. Similar stage from a safranin-gentian preparation.

247-248. First spermatocyte, condensation of chromatin granules to form tetrads in the linin spireme.

249. Later tetrad stage.

250-251. First spermatocytes, metaphase from mercuro-nitric material.

252-254. Similar stages from Hermann material, showing longitudinal split in both the bivalents, and the odd chromosome (x).

255. First spermatocyte, anaphase.

256. First spermatocyte, telophase.

257. First spermatocyte, daughter plates containing 11 and 12 chromosomes, respectively.

258. First spermatocyte, a and c daughter plates, each containing 11 chromosomes, x the odd chromosome at a different level (b).

259. Second spermatocyte, equatorial plates of the two classes.

260. Second spermatocyte, metaphase.

261. Second spermatocyte, anaphase.

262-263. Second spermatocyte, daughter plates of the two classes.

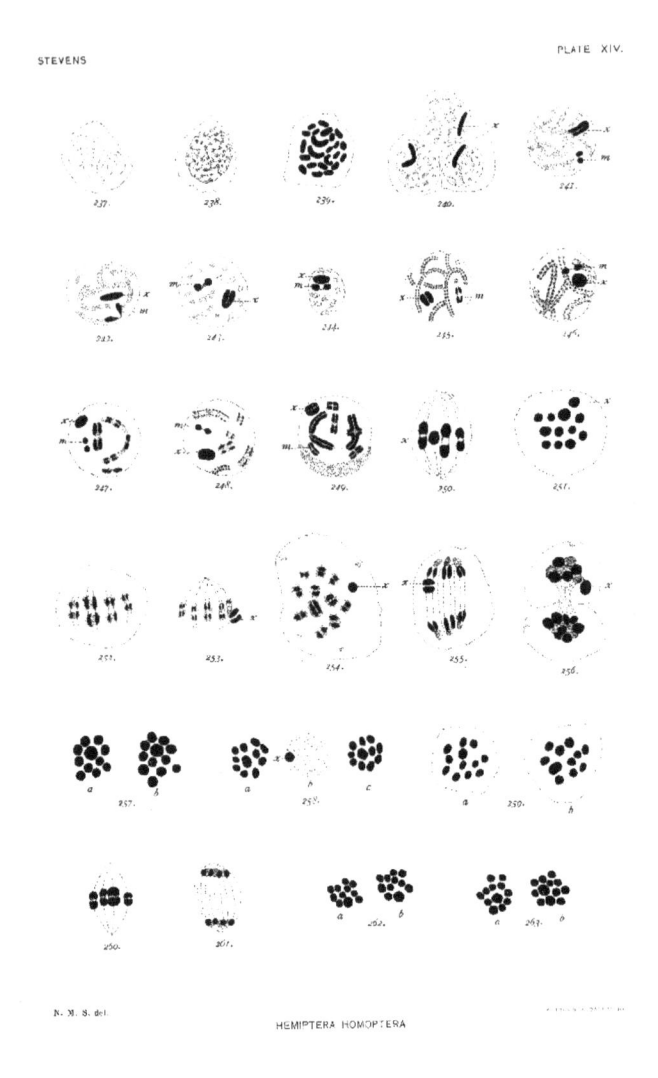

Plate XV.

Aphrophora quadrangularis.

FIGS. 264-265. Second spermatocyte, telophase, showing chromatin nucleolus (*n*) and the products of division of the odd chromosome (*x*).

266. A spermatid containing the chromatin nucleolus (*n*).

267-268. Spermatids containing both the chromatin nucleolus (*n*) and the odd chromosome (*x*), *a* the acrosome.

269. Equatorial plate from a somatic cell of a male larva, 23 chromosomes.

270. Equatorial plate of an oögonium, 24 chromosomes.

271. Resting nucleus of a young oöcyte before synapsis, showing two pairs of condensed chromosomes, corresponding in size to the *m*-chromosomes and the odd chromosome of the spermatocytes.

272-277. Sections of nuclei of oöcytes, showing one or more of these heterochromosomes, from safranin-gentian preparations.

278-279. Bouquet stage from iron-hæmatoxylin preparations, showing the heterochromosome bivalent (x).

Cacœcia cerasivorana (*Lepidoptera*).

280. First spermatocyte, synizesis stage, showing 2 condensed chromosomes (x_1 and x_2).

281-282. First spermatocyte, synapsis stage.

283-284. First spermatocyte, growth stages.

285. First spermatocyte, prophase.

286-287. First spermatocyte, later prophases, showing the heterochromosome pair (x).

288. First spermatocyte, metaphase.

289. Second spermatocyte, metaphase.

290. First spermatocyte, equatorial plate.

291-292. Second spermatocyte, equatorial plates.

Euvanessa antiopa (*Lepidoptera*).

293. First spermatocyte, equatorial plate.

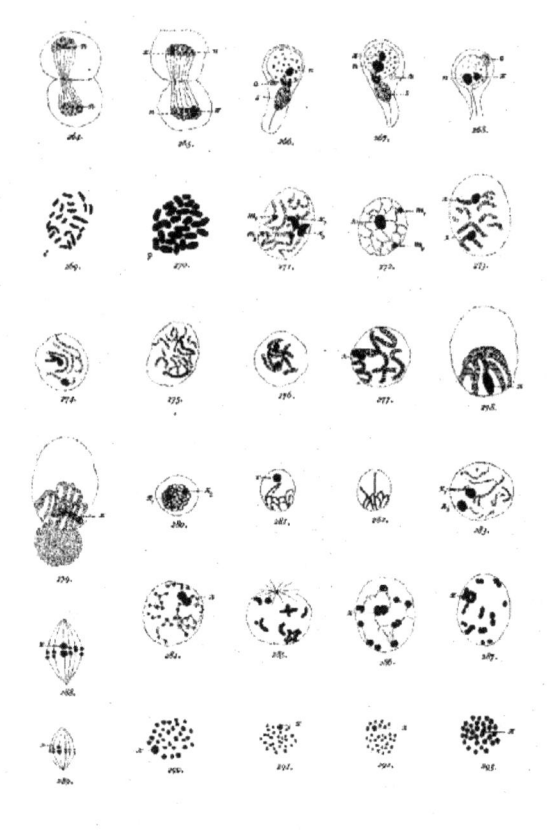

HEMIPTERA AND LEPIDOPTERA